Scandinavian Designers at Work

Jessy
Van Durme

Piet Albert
Goethals

LUSTER

Content

Introduction

'See Scandinavia and then have a baby': that was our version of 'See Naples and die'. However, we didn't just want to see Scandinavia, we wanted to experience it. We wanted a voyage of discovery that focused on our passion for the far North, photography, beautiful things in general and interior design in particular.

We therefore mapped out our journey via the homes of various young Scandinavian designers who we wanted to get to know. This book is an account of the inspiring encounters made during that journey. Of course we couldn't predict that our visits to workshops and the people we'd meet would be so fascinating, but we had dared to hope that would be the case. After all, it wasn't by pure chance that we wanted to visit these particular designers. We made a conscious decision that first and foremost we wanted to become more familiar with the young Scandinavian designers of today.

At the end of the day, so much has already been written about and exhibited from the modern generation active in the 1950s and 1960s.

We chose 17 designers of interior products and everyday products (furniture, lighting, textiles, ceramics and glassware) that piqued our curiosity. They are people with a story and a clear vision of design and life. Some of them welcomed us into their workshops, others into their homes, but in every case the personal setting was the ideal starting point for a discussion about inspiration, the hows and whys of the work process, personality, ambition and motivation.

A year ago we barely dared dream about making such a book. Meeting the designers, however, gave us the extra little push we needed to make it reality. After all, each of their projects also begins with a dream and a degree of courage.

Jessy & Piet Albert

For Ivar

Nathalie Lahdenmäki

Helsinki, **Finland**

Nathalie Lahdenmäki is a half-Finnish, half-French ceramicist who set up her shared studio with two other ceramists in an old warehouse on the outskirts of Helsinki. Her interest in ceramics began in art classes, where Nathalie got to try out lots of different disciplines. This was where she first held clay in her hands and where she soon noticed that she was often spending her breaks in the pottery studio in order to carry on working with the clay. She loved the calm and peaceful atmosphere in the pottery studio and the tactile nature of working with clay. To her amazement she later discovered that ceramics was a course at university. Her mind was quickly made up, even though she was uncertain about her professional future as a ceramicist.

During her studies she made sculptures as well as everyday objects. Her interest in everyday objects was piqued by the university campus being on the same site as the Finnish ceramics company Arabia. Kaj Franck, an icon in the world of Finnish design, was a great source of inspiration for Nathalie.

Nathalie's work has something of an austere Japanese appearance to it, a sort of reserved, simple beauty. Despite this austerity the forms are well considered. Form and proportion are for her the most important factors in a design. Functional, simple and sober but at the same time tactile, aesthetic and poetic: these different layers come together into a whole.

What she finds the most difficult is the choice - with every movement, every addition and every new form – of whether to try again. It is an intensive process of testing things out. Her work has a fragility that is achieved purely intuitively through form and colour. For her use of colour she always returns to her own colour scheme, which has remained unchanged for years. Basic colours with pastel shades that underpin them. It is very rare that a new colour crops up in a Lahdenmäki product.

RAAKAPOLTETTUJA ESINEITÄ
Näytteet
Sirpaleet 3X5
ROVASTI

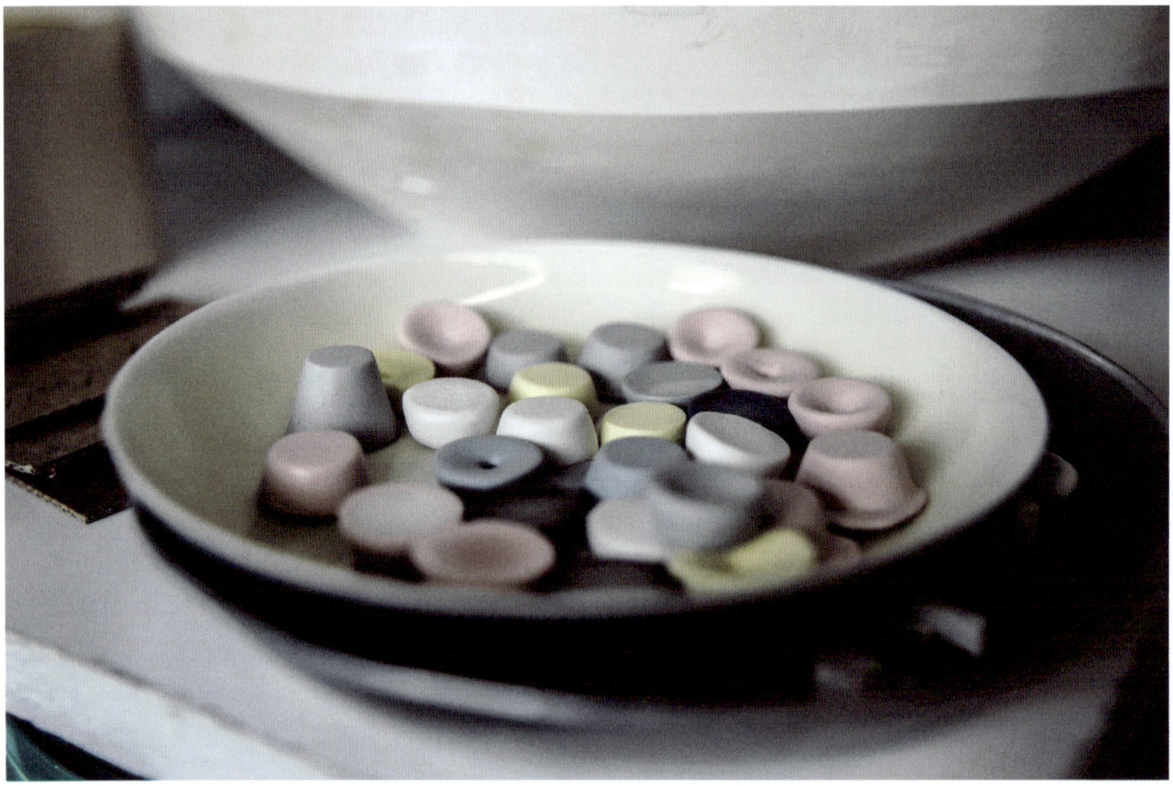

There is a clear difference between the work that Nathalie designs for manufacturers and the work she produces herself, not at least because the production process itself determines a lot. Industrial production requires a totally different way of designing than handmade production. The end products therefore have very different characters. Nathalie finds it important to identify herself with the character of the company that she is designing for. As a result, this type of work often has a much more rigorous character, with little room for chance, something more associated with handmade production. Nonetheless the work always has a simple, sober character.

However, it is her own production that is the most important part of her work. Every product is made with her hands; every object can vary or can lead to something new. Nathalie often tries out new forms, different colour combinations and textures, all of which play a really important role in the character and the experience of the object.

Besides her work in the studio, Nathalie is also a lecturer at Aalto University. The business and intensity of young people is perhaps the opposite to her calm, peaceful studio, but it is also a motivator for her. She finds it fascinating and stimulating to see young people discovering the same passion and pleasure in ceramics. The students' search for their own personality and identity also acts as a stimulant for Nathalie. It makes her aware of who she is herself as a ceramicist and of her work.

Morten & Jonas

Bergen, **Norway**

Morten and Jonas met while they were studying at the University of Bergen and after their graduation in 2011 decided to set up business together. They found a fantastic workspace on Bergen's harbour and taking a simple and informal approach, with few frills, they named their design studio 'Morten & Jonas'.

Alongside their work in the studio both designers also do work with the prison in Bergen. They support the inmates by helping them to make small-scale products. This is an important part of their life and perhaps underlines a basic idea that is at the centre of their way of working: "It's all about people."

For Morten and Jonas, the focus of design is function and the most important starting point for every design is the person or user. It is also crucial for them that their designs and methods have a positive note. Positivity, as a counterweight to all the negativity, because humour, happiness and laughter lift you up and allow you to progress. They work on the premise that you get what you give.

They consider this optimistic attitude to be the reason why they have encountered few negative experiences. Mistakes and setbacks are there to learn from and to gain experience from.

The design process for the duo is like a game of table tennis: one of them has an initial idea and then the other looks at it from a different angle, and so the ball keeps going. Sketching, model-making and prototypes are the most important tools during the development of a new product. The big driver during this process is the pleasure of creating and discovering.

As well as a positive message, many designs are also based on curiosity about a material. During the process they search for the characteristics and the boundaries of a material. Exploring and going beyond the boundaries is also an important step in the design process. Morten cites the example of a child. You know where the boundaries are and yet you still want to go beyond them.

TAKING
NYHET

Sometimes you're aware of the impossible but it's an important part of the creative process and development. With every new job, they examine the challenge and try to push themselves further.

It's not just a question of the physical boundaries of a material or an area, it's also about mental boundaries. This ensures that you always keep growing, that you grow in your abilities and as a person. It is also important in order to stay balanced. Morten believes in equilibrium, giving and taking, positive and negative. And the most beautiful starting point and vision in design is perhaps the one that they adopt: it's not all about us, it's about others.

edahls
edahls
alm
alm edahls
alm edahls
Prodema
OFFECCT
WORK BOOK
BOLON
Studio1
Montana
kvadrat

Maija Puoskari

Helsinki, **Finland**

Maija Puoskari is a passionate designer who set up her studio on the outskirts of Helsinki, where she shares a space with an illustrator. Having had her hands in clay ever since she was a child and having later worked as a florist, it was perhaps no surprise that Maija followed her schooling with a desire for something creative: she chose to study ceramics and glass design through a master in applied art & design and afterwards started working as an independent designer in Helsinki.

Maija is originally from Finnish Lapland, where nature, the seasons and the weather all play an important role in daily life. The North and this strong connection with nature are in her blood, and this comes through in her work. The natural world, which she often explores with her family, is her biggest source of inspiration.

In the past Maija has often developed ceramic products, such as candlesticks and a series of ceramic tiles, where the play of light and shadow is paired with graphic textures that together make you think of a peaceful winter landscape.

Not wanting to limit herself as a designer, she is also keen to experiment with other materials. Her desire to broaden her perspective on the design landscape has prompted her for example to use glass and wood in her work. It's always materials that have a pure and simple character that she's drawn to.

In recent years she has also worked on joint projects with her husband, Tuukka Tujula, a graphic designer. In the beginning two strong minds coming from different design disciplines didn't seem to work, but over time they have developed some beautiful projects together. The results are playful experiments with textiles and glass where her husband suggests an idea and Maija turns it into something both joyful and quirky. You find this dreamy and playful character in all of Maija's designs.

This playfulness combined with her strong bond with nature creates a magical and poetic atmosphere that pervades her objects. This goes hand in hand with Maija's frivolous way of designing, although the increasing use of colour in her recent work can also be attributed to the arrival of her son.

Her design process as a product is characterised by a lot of sketching and model-making. She calls this thinking with her hands. It's a method that works well for Maija because it contributes to the tactile nature and the experience of the object, which are important factors for her. Maija finds computer graphics and other digital methods implausible ways of working because they lack this essential sense of the tangible.

Being credible, tangible and genuine are all important characteristics for Maija and, in her opinion, are the only way a designer can come up with new, unique objects. As a designer she considers it extremely important to be true to herself and her background, both factors that have a big influence on her work.

She finds it important to fully dedicate herself to things, to approach design with the right feeling, with her heart in the right place and with enormous gratitude for what has been given to her.

3LH3
MAIJA PUOSKARI
design
www.maijapuoskari.com
VESA SAMMALISTO
illustration
www.vesa-s.com

MONOCLE
MONOCLE

LAPIN KULTA
ILLUSTRATION NOW

Sophia Lithell

Malmö, **Sweden**

Sophia Lithell is a young dynamic designer based in Malmö. She works for herself in her own studio, which is situated in a gallery space in a forgotten shopping centre on the outskirts of town. She shares this lovely spacious spot with her boyfriend, a web designer; the space is also available for young (friendly) creative types to showcase their work.

She landed here after studying at the Beckmans design school and Konstfack in Stockholm. What she developed above all during her studies was a curiosity for and attraction towards the three-dimensional form. At the beginning of her studies she was encouraged to create things by hand, and from that moment on she knew this was something for her: she found the sensation of creating something on your own, with your own hands, both calming and invigorating.

With this idea at the back of her mind, she became increasingly fascinated by crafts and trades. And to this day this plays a very important role in all her designs. Sophia has a clear message that she wants to convey through using old, often Swedish customs and trades in a new context. She finds it very important that the user, as well as the maker and creator, is aware of the created object's origin and its future. Sophia finds today's mass production problematic; in contrast to this, she wants people to once again have a feeling for everyday objects. She too wants to gain this feeling, which is why whenever she starts on a new idea she thoroughly researches specific techniques, trades or materials before she has a clear idea of what the function or the end product will be. For example, Sophia took a course at Hemslöjden (Swedish Handicraft Society) in order to understand and learn the typical Swedish custom of weaving birch bark. The fact that Sophia isn't a specialist in the crafts that she practises is also part of the process. It gives her the freedom not to have to stick to the rules, but be able to experiment and allow the old craft to shine through in a new context.

Sophia does consider the function of an everyday object in her final design, but it is not her top priority. "We've already gone beyond function." If design is all about function, then we don't need to design anything more because our basic needs have been met. It is because of this stance that her objects often teeter on the edge between art and design.

Sophia wants to create works that make us aware; she wants to stand for authenticity and honesty in the product, the appearance and the materials. As her studio expands, she sets her sights even higher and hopes that her work conveys a clear intention to the user that her objects need to be cherished.

Lars Rank

Copenhagen, **Denmark**

Lars Rank is ceramicist living and working in the centre of Copenhagen. His studio is located at the back of a shop premises where he works alongside six other designers. The wonderful space is full of light and exudes the calm one expects to find in a ceramics studio.

Lars studied ceramics at the Jutland Art Academy where after his second year all disciplines were brought together into one course. Lars therefore trained more as an artist than a ceramicist. During this period, he was particularly into conceptual art and installation art.

After his studies he had the opportunity to participate in an 'artist in residence' project in the Japanese ceramics mecca of Shigaraki. A whole new world opened up to him there, and the work that he creates to this day has its roots in this period in Japan. He describes his work as the combination of Scandinavian functionalism and minimalism with Japanese wabi sabi. Where aesthetics are described as beauty in the imperfection, the impermanent and the incomplete.

He likes to let the material speak for itself, without him controlling the work too much. Of course there are rules and techniques when you're working with clay but the decorative aspect for example happens directly while you're working with the clay, not through the use of moulds or fixed templates. Everything happens intuitively and from his gut feeling.

This freedom in his creative process also often leads to new ideas and designs, for example through a chance discovery or a mistake in an earlier creation. The vases and lamps in Lars's 'stacked' series are a case in point. The designer hit upon this idea of stacking moulds after he was casually stacking together a few misshapen pieces that had gone wrong.

His way of working is all about being open to chance happenings, being ready for the surprises that the material throws at the design.

To this day Lars makes every piece in his collection in his own studio; every piece is made with his own hands. He finds this really important because he wants to be sure that every product that he delivers is of the same quality and value. It also enables him to anticipate directly the user's experience. And every product is unique as they're all handmade and the decorative combinations are endless.

Lars does dream of one day working for a producer, but it must be one who holds the same ideas and vision for the end product. Quality, simplicity and material must be the priority.

SKUTT

Design Awards 2012
BEDSTE

Kristine Five Melvær

Oslo, **Norway**

Kristine Five Melvær is a young, driven Norwegian designer who lives and works in Oslo and has set up her own studio there. Creativity played an important role in Kristine's life long before she began her studies in industrial design in Oslo. As a child, her grandmother encouraged her to play around with and invent all sorts of things. For her, this was the foundation for her creative career.

After her studies in industrial design she wanted to learn more about graphic arts. To this day, it is these two fields that are often the source of her designs. When she's designing a product she looks for the relationship between graphics and the object. Both are about visual communication.

What attracts her to these disciplines is the combination of the freedom of the process and the way in which fixed rules or physical limits define the solution. It's always a search to get the best possible outcome and to explore the interplay between the graphics and the object.

The communicative aspect of this is particularly important. A product needs to communicate with the user. It's about the experience, the feeling and the connection between both. It's by approaching design from this angle that Kristine tries to give her products a human character. The human aspect is very important in her work.

Ideas often emerge from everyday life, daily rituals and customs. This is why it's essential that the process of developing a project is physical. Kristine finds it crucial that as a designer she comes into contact with the tactile nature of design, that the object she is designing is first of all felt and experienced by the designer.

In this respect using the right materials is key. These contribute to the story that the object tells, the image it projects.

Textiles are one material that really pique Kristine's curiosity. For her it's the material that best unites everyday objects and graphic design. However, the material that Kristine most commonly uses is glass. She finds it mystical, bright and bearing a sort of classical beauty. Glass requires care and caution; once it's broken it can never be restored to its original state, but it can last for centuries. Kristine tries to saturate her objects with this timelessness and value. She hopes to create objects that are timeless, that last and that the user cherishes.

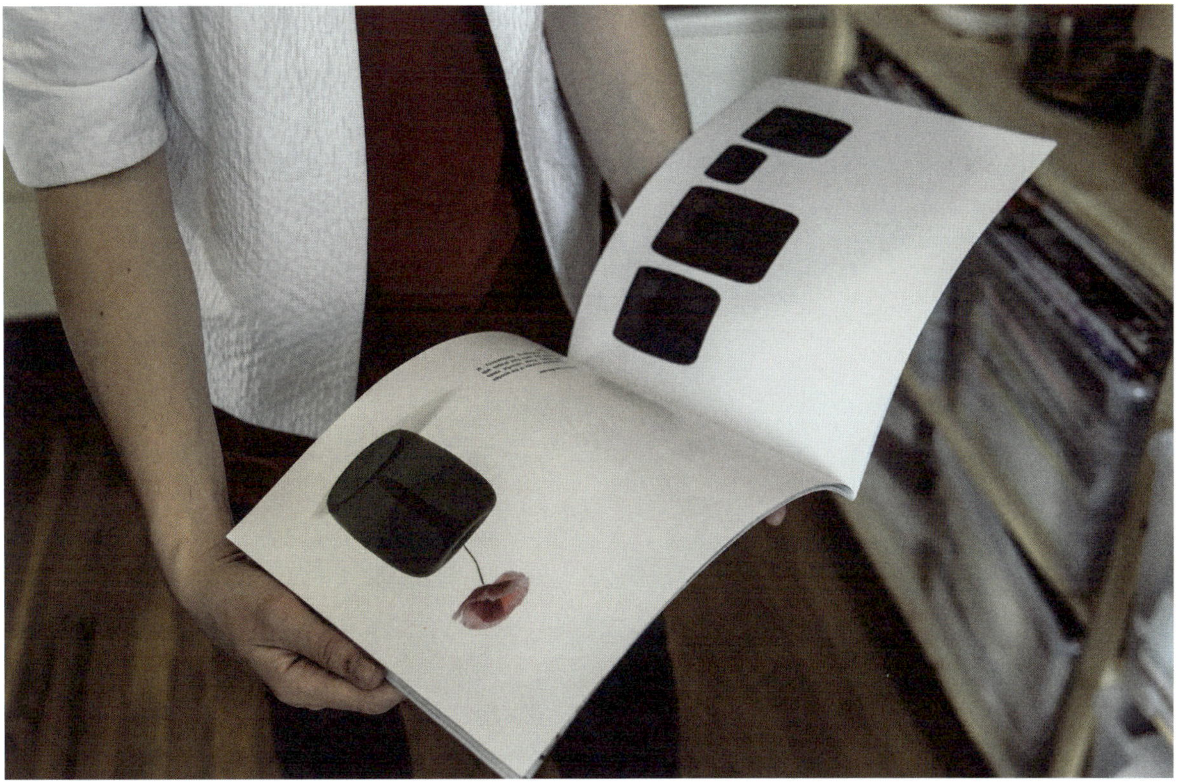

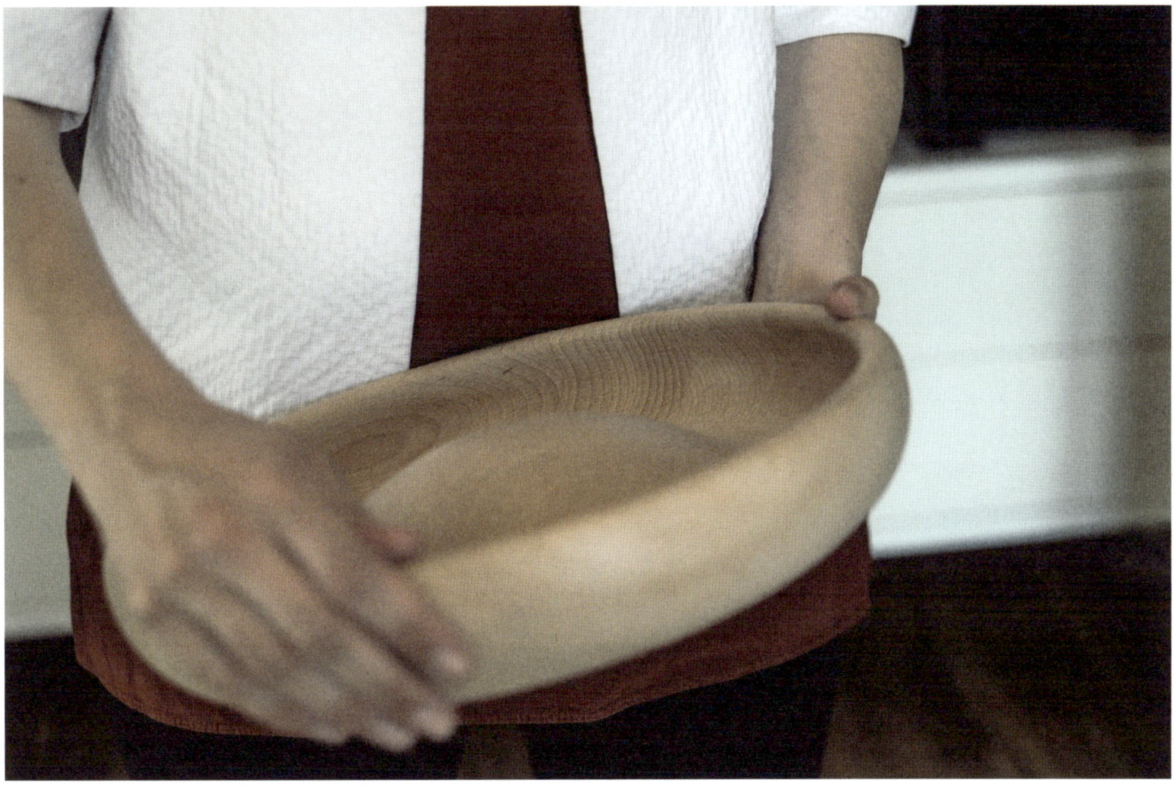

PACKAGE DESIGN
SAGMEISTER
FURNISH
PATTERNS
Barber Osgerby
CARL HIAASEN

Knudsen Berg Hindenes

Bergen, **Norway**

What started out as Knudsen Hindenes later became Knudsen Berg Hindenes, the idea being that three is a perfect number and reflects completeness. With two people you can have two-way dialogue but with three it becomes three-dimensional. With this in mind, a partnership was established between Steinar Hindenes, who had already enjoyed years of experience in design and as a lecturer at the Bergen Academy of Art and Design (KHiB), Petter Knudsen, the technical brain, and Anders Berg, who had years of experience working with wood and later became a student of Steinar at KHiB. Each of them brought their own individual strengths and insights.

As a collective of diverse personalities it is crucial for them that, despite their differences in backgrounds and disciplines, their minds are all attuned to each other's. Of greatest importance is that they all derive pleasure from designing and that each of them is fulfilled by their work. Bringing all these aspects together means that they develop their designs very quickly and efficiently.

The designs created at KBH clearly have their own language, line, material and use of colour. The design studio wants to be associated with a simple and human design language, one that's on a human scale and relates to the human body you're designing for. The designs developed here, from idea to first sketch to prototype, are not complex or sharp objects but rather soft and pure designs.

Every KBH design is fully worked through until it is almost a complete, useable prototype. This is a very important stage for KBH and one with which they want to distinguish themselves. Every model is fully tested and made on site in the studio so that the producer receives an almost finished product as a model for production.

The physical feeling of an object is very important. Being able to touch and personally experience an object is central to what KBH does. Feeling, touching and understanding a design is a very important step in the process and one to which KBH devotes a lot of time, energy and space. It also prompted them to move the studio to a bigger site where the workshop can take its rightful place and where all prototypes can be properly made, tested and fine-tuned.

The purity of KBH's designs can also be attributed to the simple and honest use of materials. The influence of materials is very important. The trio are always looking for news ways to use materials or materials from different fields in order to find a solution to the question posed by the object. They consider the use of unusual materials a challenge, a driver in order to get the most out of something. Plunging into a new material arouses curiosity; it gets you thinking and makes the designer question things.

Above all though it is the dedication, passion and love for their profession that is the force and the greatest driver behind KBH. This emanates from their workplace, their designs and their personalities. They completely surrender themselves to seeking out, drawing and designing the products of tomorrow.

formlabs

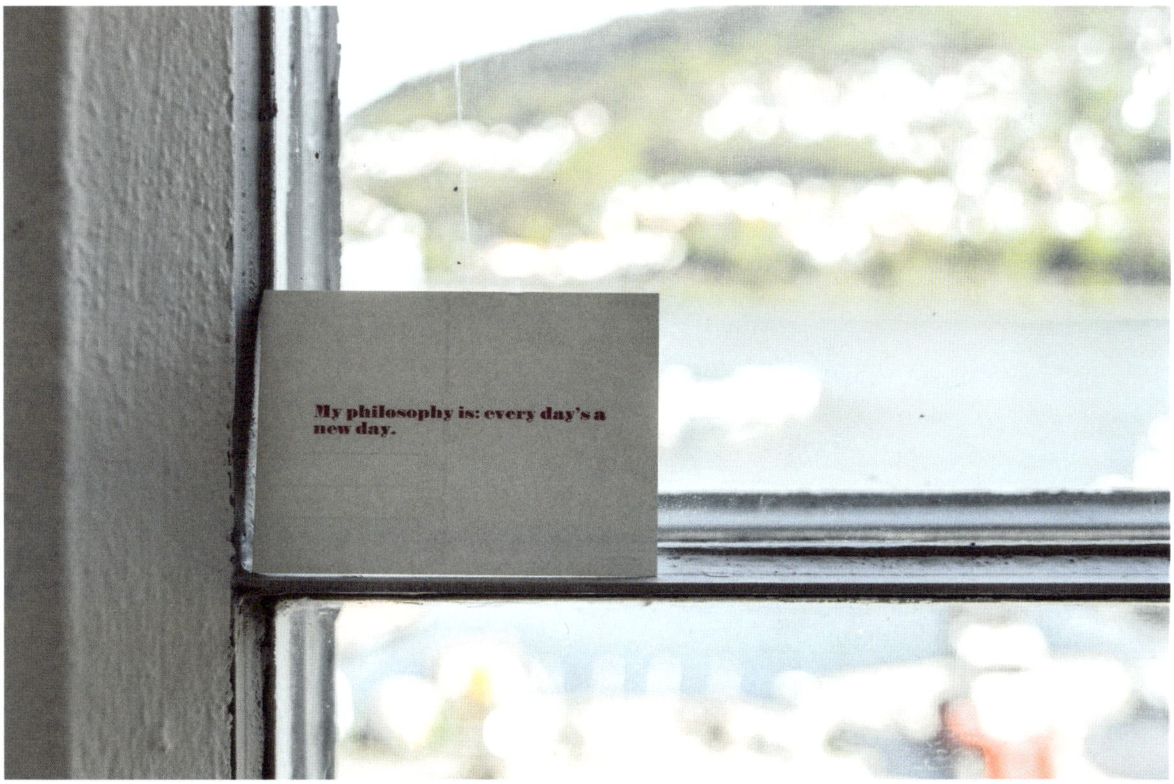
My philosophy is: every day's a new day.

Form Us With Love

Stockholm, **Sweden**

Form Us With Love design studio was founded straight after the graduation of its three creators: Jonas, John and Petrus. They had all studied together and it was during their student days that the idea started to take form, setting up a studio that shared their common vision of design. The three totally different personalities opened their design studio in 2005. This achievement was possible, they say, because each of them was striving towards the same goal, in their own way.

Form Us With Love don't consider themselves simply as product designers; they are shaping companies' brands through design. The three founders want to give existing companies a new lease of life, not just through products but also by positioning their brands.

When the studio was set up, there was little strategy about how to develop Form Us With Love beyond having a common vision, focus on the same goal and a whole lot of ambition. As for the name, this was simply the name they had always used when presenting their work at college. They checked to see if the domain name was still available. It was... and so Form Us With Love was born. The different levels of meaning in the name hint at the passion and ideas that the three designers want to share with their clients.

FUWL very quickly gained international recognition. The founders trace this is down to their shared ambition, their curiosity and above all their lack of fear. Being able to make mistakes is something they find important both for the design process as well as the studio's development. "You learn so much through your mistakes and the experience always contributes to the future", they say.

Their designs are part of a bigger picture. When they work with a client, they always start by carrying out a broad study audit of the brand; it's market, competition, audiences etc. It's all part of the designers' efforts to immerse themselves in their client's business. When it comes to the creative concept though, it's important for them to take a step back and distance themselves. They go and look for other images, break free from the standard and conventional and above all remove themselves from the world of business.

When they're developing a creative concept, they often have an extensive brainstorming session, bringing along their sketches and models and then messing about with them and puzzling over them. During the process, you always have to be prepared as a designer to question your work. It is important to allow space for discussion and be aware that there is not just one solution. This questioning of solutions and searching for the unconventional is reflected in the choice of materials: materials from a different context can provide an answer to the question raised by the product.

FUWL has already earned a great reputation, but the team is certainly not resting on its laurels. The studio wants to continue to design for themselves and for companies. Above all they don't want to stand still. They want to explore new disciplines beyond furniture and lighting and broaden the scope of their work. They see themselves as strategic designers without wanting to limit themselves to one field. They want to be open to new opportunities and be like a chameleon that can adapt its colour to each new task.

DESIGNPRIS
2007
DESIGNPRIS
2011
DESIGNPRIS
2013
FORM-
PRIS
2012
DESIGN
AWARDS
2013
2011
FORM
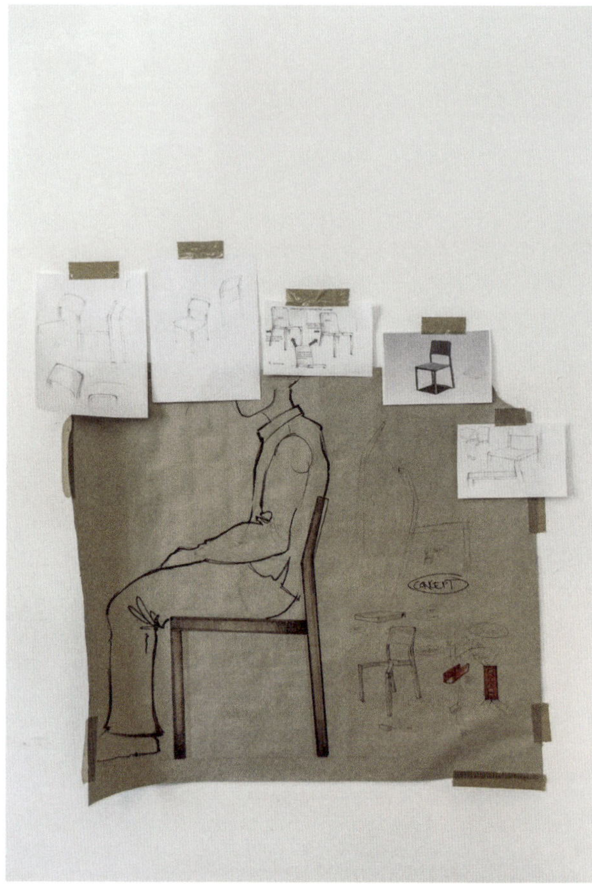

Wallpaper*
DESIGN AWARDS 2013
Best trend
'Slab' vase,
by Form Us With Love, for Cosentino/Silestone
TAMPOPO
Japan 1985
FUWL CINEMA 15.04

FORM
HAPPY
NEW
YEAR
2014
Guys!
let's make stuff
together! great meeting
you! XO B.
Bradford Shellhammer
VERNISSAGEINBJUDAN
With a note
"Sweden"
Country
manager
Jonas
Petersson
(Must have)
SAMSUNG

MARKT
HALLE
NEUN
EISENBAHNSTRASSE

Siren Elise Wilhelmsen

Bergen, **Norway**

Siren Elise Wilhelmsen is a Norwegian designer who works in Bergen in Norway as well as Berlin, the German city where she lost her heart long before she was even aware of it herself.

When she went to college, it didn't seem so much a conscious choice as a natural step to study in Berlin. Initially she planned to do an intensive course in fashion, but during her first year in the fashion academy she discovered product design and the many different materials that could be used for both fashion and products. She realised that she was drawn not just to textiles but also crafts; she enjoyed creating beautiful things from materials by hand.

Splitting her life between Germany and Norway means she doesn't fit in 100% in either place. In Berlin she is always the Norwegian, and in Norway she doesn't totally conform to the Norwegian way of working. There are of course linguistic differences in the way of expressing, but according to Siren, it is partly because of the different approaches and mentality used in Germany and Norway.

In Berlin she was trained to be a designer with a focus on conceptual designs. There was a lot of emphasis on the idea and the thoughts behind the design, which led to her experimental way of designing and a philosophical and poetic approach to objects. The Norwegian way, in her opinion, starts from a more rational standpoint with the focus on the object.

Siren often questions the concept of 'design' in her work: what is design? what makes something special? For her, design is not simply the re-styling of an existing function, but more importantly a questioning of needs, functions and existing forms. Siren doesn't want to take everything as absolute or self-evident. Her design '365 Knitting Clock', for example, questions time. Time is something that was invented to give people a stronger hold over a changing situation. Any given time is relative, measured by us in numbers and figures. With her design of the expanding clock, Siren didn't want to focus on the function of indexing time but rather point to what we as people make of time. The work could have the subtitle: 'How slow time can be'. This question is also emphasised and embodied through the slow nature of the craft, the purity and simplicity of knitting a scarf.

Crafts and handicrafts are a great inspiration for Siren and an important part of the design process. As someone who is filled with curiosity and a desire to learn, she is repeatedly fascinated by a material or technique that she then has to fully immerse herself in. It is from the knowledge gained and the techniques learned that ideas then flow. She will never be a specialist in one technique; instead she sees herself as a jack-of-all-trades.

Siren can become extremely fascinated by things or images that cross her path and then wants to unravel them. This is how her latest collection came about, influenced by Norwegian mythology. She became so fascinated by the Fantoft Stave Church near Bergen that her entire collection was based on the colours, forms and images from the immediate surroundings.

To date none of her products are available on the market. This is partly to do with the intensive process of each design and her personal commitment to it. She loves it when the idea behind the product and the feel of the product are just right. Siren doesn't want to go commercial and make compromises in her work.

In the future she hopes that her designs will end up in good hands, being produced by people who understand her ideas and vision and who place as much importance on these values as she does. She hopes that her products will in this way become valuable both in the material and immaterial sense.

STAPLES
copy
system 3
PAL
Digital Jacquard Design
Hella Jongerius
Misfit

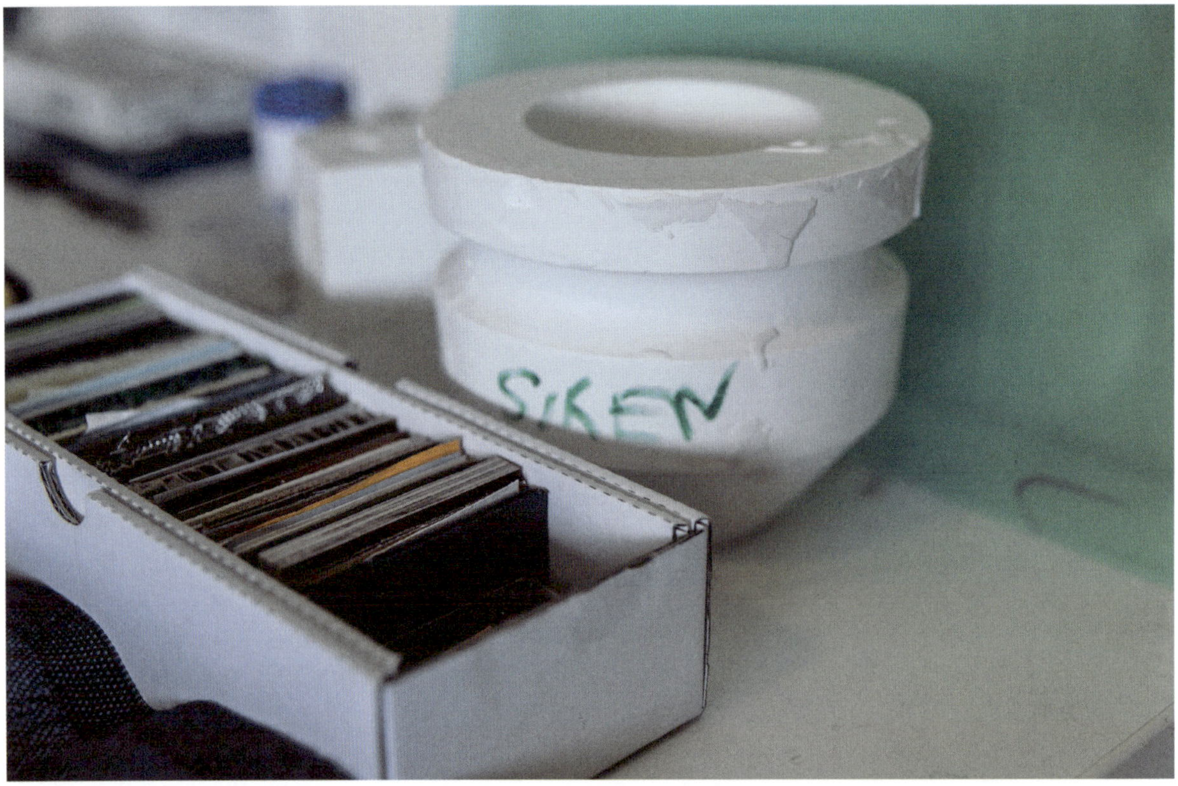
SIKEN

SIREN ELISE WILHELMSEN
DESIGN STUDIO

FELT. PURE WOOL.

Grandfather Clock
Limited Edition

The Image of the City
MUTATIONS
REM KOOLHAAS
HARVARD PROJECT ON THE CITY
STEFANO BOERI
MULTIPLICITY
SANFORD KWINTER
NADIA TAZI
HANS ULRICH OBRIST
ACTAR
THE FUNCTION OF FORM
THE MODERNIST CITY
Chicago
dictionary of advanced architecture
INSURGENT CITIZENSHIP
Ornament and Crime
Ariadne
OF THE CONFLICT
ISRAEL
PALESTINE
OPERATIVE DESIGN : A Catalogue of Spatial Verbs
ECOLOGICAL URBANISM
S,M,L,XL
THE MONACELLI PRESS
SIZA VIEIRA
LOFT
SANFORD KWINTER
FAR FROM EQUILIBRIUM
Nasjonale turistveger
Situationist International ANTHOLOGY
REVISED AND EXPANDED EDITION
Bureau of Public Secrets
LEARNING FROM LAS VEGAS
visual aid²
visual aid
EDUARDO SOUTO DE MOURA
LOFT
Deleuze &Guattari
A THOUSAND PLATEAUS
Visual Complexity
Mapping Patterns of Information
CONVERTED ARCHITECTURE AND TRANSFORMED BUILDINGS
Build-On
gestalten
ARCH+ 196/197
POST-OIL CITY
I swear I use no art at all. 10 years, 100 books, 18,788 pages of book design by Joost Grootens
BIONIK & DESIGN
VISUALISING INFORMATION IN GRAPHIC DESIGN
form 228
Inspira

Simon Key Bertman

Stockholm, **Sweden**

Simon Key Bertman works as an independent textile designer. Studying textile design felt to him like the natural choice, as did his decision immediately after his studies to set up his own design studio.

What wasn't so obvious was his decision to start up commercial production on his own. His studies had focused primarily on artistic creations, so commercial textile design was not a given. The recent graduate needed courage, persistence and guts in order to bring a collection to market so soon and all on his own.

Still today, Simon's own production line takes up most of his time. That said, he also makes a few personal and artistic works. It had never occurred to him to work for other companies as well, but this is something he plans to start doing in the near future to broaden his horizons and gain new insights.

As for his actual designing, Simon always remains faithful to one of two methods. Both approaches are based on the relationship between the surface, the pattern and the use of colour. He combines this with a mathematical approach for generating the pattern, and he also places great emphasis on the tactile nature of the finished product.

Using one of his methods, he seeks to explore the boundaries of the object and to play with this using an interesting pattern. For this, he always works with one or two colours or shades in order to get the best effect of the interplay between pattern and surface.

For the other method, Simon sometimes chooses a basic pattern, one that is easy to understand, and then adds a complex use of colour so that the simple pattern gains an extra layer.

Simon finds it important that his designs are simple and understandable, characteristics that make most of his work accessible. And his use of pure and classic materials such as wool, cotton or linen makes his creations durable seem timeless.

12

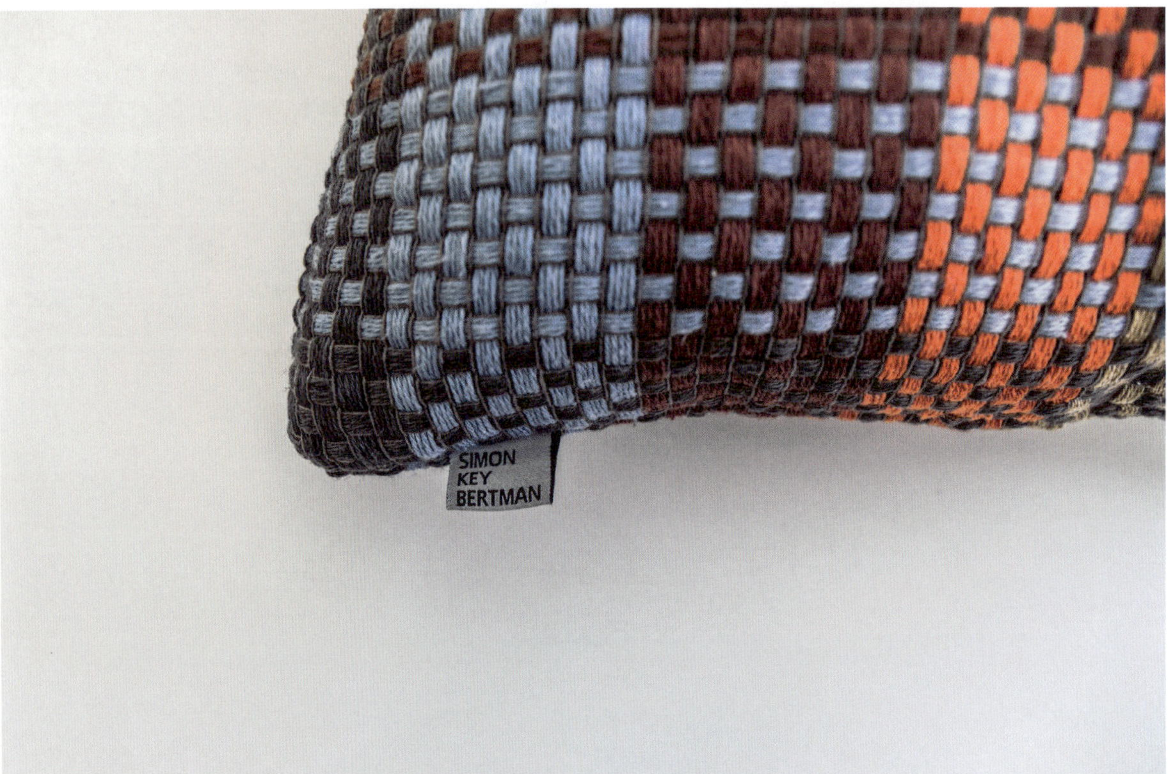
SIMON
KEY
BERTMAN

P
PILOT
Fineliner

Kristine Bjaadal

Oslo, **Norway**

Kristine Bjaadal studied design at Oslo National Academy of the Arts. After her Bachelors it was clear that her interest lay with the detailed work of small objects. For her Masters she therefore focused on ceramics and textiles.

To this day the details and the feel of small everyday objects remain the biggest driver behind the designs that she makes. It is also vital for her that the small everyday objects are made on a human, tangible and real scale. They are manageable and perhaps even practical to design.

Kristine loves to accumulate things. This collecting and cherishing of small objects triggered her interest in objects in which you can store things. A lot of her works reference this by exploring the relationship between the open and closed, covering and discovering. She often tries to make this aspect of her work easy to read.

To explain this 'readability' as well as the sentimental character of her designs, she refers to the walnut, a reoccurring source of inspiration for Kristine's everyday objects. You can deduce from the form and the characteristics of the walnut that it is protecting something valuable, that there is something behind the shell that makes it worth cherishing and taking caring of.

At the start of the design process Kristine usually has a clear picture in her mind of the feeling or gesture that the object must convey to the user. Exactly how that evolves into the design and the result is determined by the process itself. Kristine very quickly starts to turn the shape and the idea into a model in order to get a clear feeling for the object before she then further develops it into the end product. For Kristine, the most difficult part of the whole process is taking decisions as deciding means making choices.

Kristine is not a specialist in any one area, but instead has a general knowledge of various disciplines. She is aware not only of the difficulties but also the positives that this can bring. She believes that it makes her less restricted by a material or a technique and often able to gain new insights totally by chance.

What Kristine wants to achieve above all else is that her designs get people thinking, even if indirectly, just like she starts thinking about the indirect and accidental directions that she takes, never knowing where she'll end up but nonetheless always being surprised and driven forwards.

Katriina Nuutinen

Joensuu, **Finland**

Katriina Nuutinen lives and works in Joensuu in eastern Finland. She has a workspace in an old wooden school in Utra, an old living and industrial area in the suburb of the Joensuu city; it's the place where she finds peace and inspiration.

Katriina designs everyday objects that often look a little bit different than the archetypal product, more conceptual. This conceptual approach is a legacy of her studies at Aalto University in Helsinki, which encourages and develops this way of thinking. Katriina doesn't want her everyday objects simply to have a functional role; she wants to tell a story through her work. A lamp is more than an object that emits light; it can convey an entire story or meaning to the user.

Katriina's designs don't usually originate from a commission, but rather from a simple idea. The idea could come from a previous piece of research or a project, an old sketch, photos or images or even a conversation with a friend. Any of these could be the trigger for a new piece.

Katriina always tries to find a balance and an emotional tension in her work. For her, a design is powerful and is made up of a story as well as a shape, colour, composition and the use of materials. Her pieces never follow trends or fashions; they emerge purely intuitively from a feeling or a personal influence of a particular moment. Katriina doesn't want to design trendy products, but wants to stand for strong, timeless designs that retain their value over the years.

Design les

Design less, think more

sign less. think more

Katriina has a clear set of values that consciously guides her designs. Her products radiate fragility and emotion. They convey both strength and vulnerability. These contrasts are often connected with the carefully chosen materials that she uses in her work. She consciously chooses valuable materials such as wood, glass, ceramics, leather and steel, in other words materials that have proved their value and above all are natural materials.

Katriina takes a craftsmanship approach to working with these materials. This approach derives partly from her own experiences during her ceramics and glass studies and partly out of respect for integrity and quality production. Such craftsmanship gives an extra dimension to an object. It is handled by human hands and contributes to the fragility, the uniqueness and the value of a product.

Susanna Vento

Helsinki, **Finland**

Susanna Vento is an interior designer and stylist who lives and works in her apartment in the centre of Helsinki. After studying interior design she began working as a stylist for various magazines, exhibitions and books. But alongside this passion for styling and organising, there was always this urge to design and create.

One result of this inner drive has been the development of the Varpunen sack, the 'prototype' bag for storage. One day when Susanna was at her husband Jussi's printing office she saw some plain advertising sheets used for billboards and realised it would make an ideal material for storing her eldest daughter's bath toys. Susanna's background in styling and her interest in graphic design resulted in prints being added to the polyester sheets, and the end result was the first Varpunen sack.

It is this alternative look at a material and what it can be used for that has led to the development of new products.

What began as something the size of a children's toy grew into a basic version the size of a laundry basket. The general design of the object means that the user is free to decide what he or she personally wants to use it for, either giving it a fixed purpose or adapting to the needs of the moment.

This fascination with graphic design, lines and composition is not only found in the Varpunen sacks; Susanna's DIY book, Omin Käsin, published in 2013, is also full of similar inspirations.

The book, a collection of her own creations, is a means for Susanna to openly share her work and encourage other people to create in a simple way. Each of the designs in the book radiates the same simplicity and logic, strengthened with the same pure graphic design. Susanna is happy that she is able to throw her heart and soul into what she's good at and enjoys sharing her ideas with others in order to make her environment that little bit more beautiful.

SCHOLTEN & BAIJINGS
ÄITI
ISI
OLEMME LAIVAMATKALLA
VILLEVIKINKI LAIVALLA
OIKEASTI!

TACK!

BLACK FLAG
DARFÜ

TYPE PLAYER
STUDIO VISIT

BRIO.

SMEG

FINNISH DESIGN
DEKON SISUSTAMA
ASUNTOMESSUTALO
RAJOJA
204
SIVUA!
KESÄ
ok
omin käsin

Mika Tolvanen

Helsinki, **Finland**

After his studies at The Royal School of Art in London, Mika had the opportunity to work with designers including Tom Dixon. After a few years, however, he returned to Finland to set up his own studio in a collective working space in an old coffee shop. Working in this collaborative way brings an added value, in Mika's opinion. There is no direct input into each other's work, but it's a supportive atmosphere with feedback available. He considers this really important and it's a large part of the creative process.

For Mika, another essential part of creating a design is sketching. He often begins with a few ideas, details or hastily drawn lines; these rough ideas can sometimes begin as a chair and then transform into a lamp or a table. He owns dozens of folders full of rough sketches that he has drawn over the last 10 years. These can often be a source of inspiration when he's asked to do a new assignment. A mere detail can provide him with inspiration for a design and give rise to a totally new idea.

It is his initial sketches and the subsequent 3D plans that Mika uses to make his first models. He always makes these full scale because of the importance of ensuring that the design's shape and its impact in the space are right. Even though the models are often made out of foam, card or wood (and therefore not usable as a final product) they are one of the most essential elements for Mika to realise his final design.

Even things like ergonomics are often tested on these models so that form and function are perfectly matched. For Mika it is crucial that products fulfil their function, and he seeks a form that provides a perfect aesthetic interpretation of the product's function.

Mika likes assignments that are clear-cut, where he has one goal and where he can logically work towards that goal. His approach to design is above all a rational one. He is interested in conceptual design, but it's not in his blood. Instead he finds himself drawn to a sober and simple looking product.

The language that pervades his designs comes about intuitively as he's making the models and prototypes. He trusts his sense of proportions, simplicity and form. Mika is a quiet and reserved person, characteristics that also come through in his work. A design by Mika will never scream for attention, but it will always make a deliberate and strong statement.

As a designer, he's continuously setting himself challenges and is therefore often working on several designs at the same time. This ensures that he never tires of a project and gives him extra motivation to finish each design perfectly. For him that means stop when you've done enough or continue until just the essence remains.

ETAPPI

Microsoft

SPEEDRY
Magic Marker
Twin Tip

Jonas Wagell

Stockholm, **Sweden**

Jonas studied interior architecture and furniture design at Konstfack University College in Stockholm. After completing his studies his first job was at an architecture firm. During this period, he also carried on designing objects and furniture, and after a while he decided to take the plunge and start working full-time as an independent designer. He set up his studio in an old milk shop in the lively district of Södermalm in the Swedish capital.

It was a step into the unknown, but this didn't deter him. Nothing ventured, nothing gained, as the saying goes.

Jonas works as both an architect and a product designer. Sometimes the focus is more on one discipline, sometimes the other. However, focus is one of his basic requirements for a productive way of working.

Each part of his work requires its own approach. When he's working with architecture it's often based on a concept. For Jonas, this is like solving a jigsaw puzzle, where the individual pieces are the different aspects of a project. This type of work is about creating an experience of a space and solving functional requirements. He's drawn to clever solutions, but simplistic aesthetics. Designing product is based on the same principles, but still quite different, Wagell says. He tries to create intuitive objects and simple and clear forms, but with a certain character highlighted or emphasized. Consumer products need to communicate concept and idea in a matter of seconds, he concludes.

The product in the industrial context is something that fascinates Jonas; designing an object that goes into production and through the hands of skilled engineers and refined machinery is something he finds poetic and exciting. Essentially, mass-production consists of detailed industrial works much like arts and crafts.

Jonas believes it is important to develop a clear direction and a personal language. This enables a designer to create his own identity and produce work that has distinct character, but still is developing and innovative. As a designer you are your own brand and need to create your own recognition, he says.

He calls himself a minimalist, not so much in the sense of straight lines and austere forms, but rather through the omission of the superfluous. If the answer is within your reach, don't go searching for complicated solutions. He prefers to find simple and analogue, although not obvious, ways to solve problems.

JOY

HOMAGE TO A TYPEFACE

COOKIE JAR

Jonas is a happy and energetic person. He is inspired and impassioned by everyday things, travels and above all a good sense of humour. This comes across in his rounded and cheerful designs, often finished off with a distinctive colour. According to Jonas, colour creates attraction and is a tool for communication. However, all products need to work in black and white as well.

The cheerful designs are, despite their roundedness, very often geometric and graphic. Jonas strives to make 'icons' out of these refined and pure forms, with their particular use of colour and individual style. He seeks images that linger in the memory, shapes that we will remember.

Developing his signature design has been a long journey. A strict attention to detail and lots of time, effort and patience are the principles that have guided him along the way.

Being a designer is something very personal, you need to remain passionate and interested and always be open to learning. And remember to have fun along the way, Jonas adds with a smile.

WORK
BONYTT NR 13/2011
ACKURAT
DIZAJN PARK
MODERN DECORATION HOME
FRAME 22
FRAME 26
FRAME 35
MARU
NEUTELINGS RIEDIJK ON TV
DESIGN DIFFUSION NEWS 152
KONTORSKATALOG
Wallpaper*

HOMAGE TO A TYPEFACE
1000 chairs
ARK
VISUAL STORYTELLING
Out of the Box!
SUBLIME
FRAME
PARTY PEOPLE
RESTAURANTS STANDS BANK PATRICIA URQUIOLA STUDENT WORK
mission

Sami Kallio

Gothenburg, **Sweden**

Sami Kallio was born in Finland but grew up in Sweden. After training to be a carpenter, he was curious to learn more about design. So he started studying design at the Academy for Fine Arts in Gothenburg and then opened his own studio not far away.

His Finnish roots are nonetheless very important to him, and it is interesting to note that the first collection he presented to the public was entitled 'The Finnish Blood in Me'. Sami finds having both nationalities an enriching experience. His designs reflect this, having both a note of Finnish melancholy and a touch of Swedish tradition.

The inspiration for his objects comes from everyday things. Sometimes this can derive from a need of his own, sometimes from a detail or a technique that he researches further and then tries to interpret. Developing a product starts with a rough sketch that Sami later seeks to produce digitally. However, the most important part of the object's development is the phase in the workshop. It is here, in an annexe to the office space, that Sami spends most of his time. It usually takes two to three models before he creates a prototype. Sami finds it very important to work manually on his objects and he tries out all possible ideas and details on a 1:1 scale.

His training as a craftsman has a big impact on his way of working. At any one moment, he can combine the knowledge and technical expertise he has gained as a carpenter with the creative and innovative thinking of a designer. Given his background as a carpenter it is little surprise that wood is the material he works with most, enjoying its familiarity and how it takes shape in his hands.

Another important aspect of Sami's designs is the inclusion of colour. He doesn't always find it essential, but colour is able to strengthen an object or place it in another context. Colour is like a detail for him: it can either strengthen or weaken a product.

Sami wants his designs to be a trigger that gets their users thinking. He wants his designs to surprise, not by having a mad shape but through a subtle deviation or a new but authentic way of creating. It is precisely the uniqueness of an object and its effect on the user that makes a design timeless and enduring.

KLASSIKER DES 20. JAHRH
DERTS
SCHÖNER WOHNEN

1900
-1949
50
60
70
80
90
DIE KLASSIKER

Louise Hederström

Malmö, **Sweden**

Louise Hederström just moved together with her graphic designer husband to a splendid apartment in Malmö which is still developing and where life and work intermingle. She finds this the perfect way to stimulate her productivity.

Ever since her childhood, Louise has felt an urge to create. Growing up on a farm she could experiment to her heart's content with all the materials that she found in her surroundings. This love for creating has stayed with her into adulthood.

Despite a great interest in art, Louise went to study at the Beckmans Collage of Design in Stockholm. She graduated in 1998 and began a career in design. She was eager to broaden her knowledge and develop her skills by working at various design studios rather than immediately setting up on her own. Louise's career has been a marathon more than a sprint, working slowly but surely towards her goal. In 2000 she set up her own studio to step by step let it grow to what the studio presents today.

For Louise, design should stand out from the crowd by having personality. She sees each of her designs as a unique individual with its own character, and yet never loses sight of its functionality.

Materials play a very important role for her. They are not simply the means with which to realise an idea; it is often the material itself that gives rise to an object, that invites and encourages a particular means of expression or a particular solution. Hederström considers it important to listen to her materials so that she can fully understand them.

Metal is one of the materials she uses the most in her designs. Metal is, for Louise, a very honest material: it doesn't have any crazy laws or restrictions, it is easy to understand and has great purity. She even suspects that she repeatedly used the metal earlier because it was a safe choice; she knew the material and knew how to handle it. And this characterises Louise as a person. She will never take an uncalculated risk but rather, just like the build-up to her own studio, she goes looking for the right answer step by step.

Through her designs Louise also wants to stimulate our awareness about how we handle the objects that we use. For example in the pouf 'LAYER' she designed a basic piece of furniture out of fabric that can be sat on or leant against in various ways, but the user must fill the pouf him- or herself using old clothes or materials etc. It's a principle from times gone by, and with this design Hederström wanted to raise the question of why we no longer re-use things as much.

A second aspect that comes to the fore in this design philosophy is encouraging the user to identify with the everyday object. The designer leaves space for users to inject their own ideas. And this user-object relationship then creates a personal bond between them once again, giving the product an emotional charge.

Looking to the future, Louise hopes to further develop her career as a female designer working with good manufacturers who share the same concerns for design - and with a bit of output on the international market.

Jessy Van Durme (1988) is an interior architect and furniture designer who loves being inspired by contemporary Scandinavian aesthetics and vintage design from the far North.

Piet Albert Goethals (1981) is a photographer – his work focuses on travel and music. During the last 15 years he has regularly travelled through the Nordic region, taking portraits and landscape shots.

Nathalie Lahdenmäki
nathalie.lahdenmaki@aalto.fi
www.nathalielahdenmaki.fi

Morten & Jonas
post@morten-jonas.no
www.morten-jonas.no

Maija Puoskari
maija@maijapuoskari.com
www.maijapuoskari.com

Sophia Lithell
sophia@sophialithell.com
www.sophialithell.com

Lars Rank
Lars Rank Keramik
lars@rank.dk
www.rank.dk

Kristine Five Melvær
me@kristinefivemelvaer.com
www.kristinefivemelvaer.com

Knudsen Berg Hindenes
hello@kb-h.no
www.kb-h.no

FUWL/Form us with love
info@formuswithlove.se
www.formuswithlove.se

Siren Elise Wilhelmsen
post@sirenelisewilhelmsen.com
www.sirenelisewilhelmsen.com

Simon Key Bertman
Simon Key Bertman Textile
Design & Art
simon@bertman.nu
www.bertman.nu

Kristine Bjaadal
kristine@kristinebjaadal.no
www.kristinebjaadal.no

Katriina Nuutinen
katriina@katriinanuutinen.fi
www.katriinanuutinen.fi

Susanna Vento
Varpunen
ventosusanna@gmail.com
www.varpunen.com

Mika Tolvanen
Studio Mika Tolvanen
info@mikatolvanen.com
www.mikatolvanen.com

Jonas Wagell
JWDA - Jonas Wagell Design
& Architecture
info@jonaswagell.se
www.jonaswagell.se

Sami Kallio
Sami Kallio Studio
sami@kallio.nu
www.kallio.nu

Louise Hederström
Louise Hederström Design Studio
hello@louisehederstrom.com
www.louisehederstrom.com

Acknowledgements

We'd like to thank all the designers who contributed to this book, for the interesting discussions and the welcoming reception, almost always accompanied by a Scandinavian dose of coffee, tea and cinnamon. Thanks also to our families and friends for their belief in and support for our step into the unknown.

Colophon

Scandinavian Designers at Work

Text
Jessy Van Durme
www.studio-klaer.com

Photography
Piet Albert Goethals
www.piet-albert-goethals.com

Graphic design
Stephanie Specht
www.stephaniespecht.com

D/2014/12.005/17
ISBN 978 94 6058 1328
NUR 656

www.lusterweb.com